An *Easy* Guide to the Casio Scientific Calculator

Its functions and applications

D.M. Payne
M.Sc., C.Eng., M.I.E.E.

DP PUBLICATIONS LTD
Aldine House, Aldine Place,
142/144 Uxbridge Road,
London W12 8AW

1991

AUTHOR'S ACKNOWLEDGEMENT

The author wishes to thank **Casio Electronics Company Ltd** for their advice and assistance.

This is not an official **Casio** ***production, and any opinions in it are those of the author and not of*** **Casio Electronics Company Ltd.**

Dedicated to the memory of H.H.Payne

ISBN 1 870941 85 3

A catalogue record for this book is available from the British Library

Typeset by
Kai, 21 Sycamore Rise,
Nottingham

Printed by
The Guernsey Press Company Ltd
Vale, Guernsey, Channel Islands

CONTENTS

PREFACE

The unique capabilities of a scientific calculator can benefit everyone, from engineers and scientists, to business people and housewives. In fact, the relatively low cost of these calculators has made them widely available. Nowhere is this more evident than in the educational market, and students in particular have been leading beneficiaries of Casio's strong developmental capabilities and outstandingly innovative ideas.

Indeed, Casio's continued commitment to an aggressive programme of research and development has resulted in considerable investment in new technology and has allowed the Casio product line-up to expand rapidly. This can be seen clearly in our range of scientific calculators - a field in which we remain market leaders through the introduction of a whole host of revolutionary new products over the past 15 years.

This period of growth has seen the introduction of new ranges of scientific calculators that are more 'user-friendly' and with additional benefits at no extra cost.

Furthermore, Casio's continued commitment to the educational market worldwide has seen an increase in the range of scientific calculators that have been tailored to meet student requirements. These developments are a result of the feedback we receive from our many customers.

We have learned that many owners of Casio hand-held scientific calculators do not use these sophisticated devices to their fullest potential. In response to this, and as part of our continued commitment to our customers, we are recommending this book to you, as we believe it will help you, the reader, obtain maximum benefit from your scientific calculator.

CASIO ELECTRONICS CO LTD

INTRODUCTION

This book will enable all owners of Casio scientific calculators to understand the numerous functions of their machines and be aware of their time-saving, practical applications.

Even the least expensive scientific calculator now possesses an extremely wide range of functions, and many people find certain features baffling. Once understood, however, these sophisticated calculators will easily solve practical maths, science and engineering problems. ***An Easy Guide to the Casio Scientific Calculator*** covers the functions and uses of these calculators in a way that is comprehensible and easily applicable to the likely needs of the user. It deals with each of the various functions and concludes with worked examples to confirm understanding.

It is assumed that readers have sufficient background knowledge to be familiar with trigonometric functions (Sine, Cosine, Tangent), concepts of number bases, 2's complement numbers, and simple statistics. For the same reasons, there are no explanations of the simpler functions such as l/x.

An Easy Guide to the Casio Scientific Calculator covers all models (excluding programmable models) in the range, as there are only small differences in the function keys of the various models (see the Keyboard Guide - page vii).

Letters from readers making suggestions for improvements are always welcome.

D.M. Payne
1991

KEYBOARD GUIDE

Variations between models

There are two possible variations in the layout of your **Casio** scientific calculator:

a. ***SHIFT/INV KEY***

Where a key is marked with an alternative function, another key must usually* be pressed to bring in to play this alternative function. In some models this key is labelled SHIFT, in others it is labelled INV.

b. ***MODES***

A number of functions can only be obtained when the calculator is in the correct mode. Colour coding is also used to identify these functions.

*as long as the text on the machine identifying the function and SHIFT or INV keys is the same colour (normally brown).

The Mode Key

All modern **Casio** scientific calculators can operate in a number of different ways or Modes. To select a Mode, press [MODE] and then another key. To find which other key, look at the list printed just below the display window of your calculator.

For example, to operate in Radians Mode look along the list to find RAD. In the box next to it is a number – usually 5. So, to enter Radians Mode, press:

[MODE] [5]

The display will then show either the letter R or the word RAD.

To return to Degrees Mode, look for DEG in the list. The number 4 will usually be found next to it. You therefore enter:

[MODE] [4]

The display will then show either the letter D or the word DEG.

More information about the uses of the various modes is given in the text.

Key functions

Some keys are not included because either:

- the key is so universal that no special instructions are required. General Keys are all of this type, as are some of the **Function Keys,** the **Special Keys**, and such things as square roots, squares, powers, roots, hexadecimal number entry keys A to F etc;
- the key is really of academic interest only, eg [Σxy] (this function forms part of the process whereby the calculator does Linear Regression arithmetic).

	Key	Function	Page
Memory Keys	MR	Independent memory recall	44, 46
	Min	Independent memory in	44-45, 52-53
	M+	Memory plus	44
	M–	Memory minus	44, 46
	Kout	Constant memory recall	45, 47
	Kin	Constant memory in	45, 47
	M↔X	Memory/display exchange	44
Special Keys	[(--- ---)]	Brackets	1-3
	EXP	Exponent	3
	X↔Y	Register exchange	10-13
	X↔K	Constant memory/display exchange	45
	RND	Rounding off internal value	7-8
Base-N Keys	DEC	Decimal	18-20
	BIN	Binary	18-19
	HEX	Hexadecimal	18-20
	OCT	Octal	18-19
	AND	And	21-22
	OR	Or	22-23

	Key	Function	Page
Base-N Keys	XOR	Exclusive Or	23
	XNOR	Exclusive Nor	24
	NOT	Not	21
	NEG	Negative	20
Function Keys	hyp	Hyperbolic	40-41
	log	Common logarithm	42
	10^x	Common antilogarithm	42
	ln	Natural logarithm	42-43
	e^x	Natural antilogarithm	42
	ENG, ←ENG	Engineering	6-7
	$a^b/_c$, d/c	Fraction	48-49
	x!	Factorial	36
	R→P	Rectangular to Polar	9-12
	P→R	Polar to Rectangular	9,12-13
	Ran#	Random number	38-39
	nPr	Permutation	36-37
	nCr	Combination	37
	f p n μ m k M G T	Engineering symbols	7
Statistical Keys	KAC	Statistical register clear	27, 32-34, 45
	DATA	Data entry	26-28, 31-35
	XD	Data entry	26
	DEL	Data delete	28, 35

(cont. overleaf)

	Key	Function	Page
Statistical Keys	X_D, Y_D	Regression analysis data entry	31-35
	$y\sigma_{n-1}$ $x\sigma_{n-1}$	Sample standard deviation	27, 30
	$y\sigma_n$ $x\sigma_n$	Population standard deviation	28, 30
	$\bar{x}$ $\bar{y}$	Arithmetic mean	26-28, 30, 33-34
	n	Number of data items	27-28, 30, 34
	Σx Σy	Sum of value	26-27, 30, 33
	A	Constant term	30, 32
	B	Regression coefficient	30, 32
	r	Correlation coefficient	30, 34
	$\hat{x}$ $\hat{y}$	Estimator	30, 32-33
Complex Keys	i	Imaginary part of complex number	50-53
	arg	Angle of complex number	52
	conjg	Conjugate of complex number	–
	|z|	Modulus of complex number	52
	Re↔Im	Switch between Real and Imaginary display	51
	◄	Scroll display left	51
	►	Scroll display right	51

Chapter 1
Brackets [( )]

This chapter is intended to help you to make proper use of the bracket keys [(and)] on the calculator. By using these correctly, complicated formulae can be worked out without the need to store intermediate results in memory.

Built-in Priorities

There is a standard order in which operations are carried out on complex mathematical expressions. These rules are built into all Casio scientific calculators. The following order of precedence is used (highest priority first).

1. Expressions within brackets.
2. Functions such as *SIN*, *COS* etc.
3. Powers, *P*→*R* & *R*→*P*, Permutations and Combinations.
4. Multiplication and Division.
5. Addition and Subtraction.

You will notice that expressions in brackets head the list; *brackets are a means by which you can change the order in which expressions are worked out.* So, putting a humble addition within brackets takes it to the front of the queue.

Using Brackets Correctly

Where brackets are already included in the expression to be worked out, you just have to enter it into the calculator from left to right. However, some expressions where brackets are required are often written without them. Some common examples are shown below:

(i) $\dfrac{3.5 + 1.5}{10 - 7.5}$ (ii) $e^{3/8}$ (iii) $\sqrt{45 + 78}$

In the first case, the horizontal division line implies that the RESULT of the expression above the line is divided by the RESULT of the expression below the line, i.e. the top and bottom expressions should be in brackets.

In the second case the power of e must clearly be calculated first, so it should be enclosed in brackets.

Finally the sum under the square root sign must be worked out before taking the root, so it should be in brackets. The three expressions then become:

$$\text{(i)}\ \frac{(3.5+1.5)}{(10-7.5)} \qquad \text{(ii)}\ e^{(3/8)} \qquad \text{(iii)}\ \sqrt{(45+78)}$$

Because of difficulties with expressions such as the first, many users work out the top and bottom expressions separately, storing one result in memory before carrying out the final division. This is not the best or quickest way to get the right answer.

Here are the three expressions correctly worked out on the calculator:

Example 1

Keys	Display	Comments
(3 . 5 + 1 . 5)	5.	Result of 3.5 + 1.5
÷ (1 0 − 7 . 5)	2.5	Result of 10 – 7.5
=	2.	Answer

Example 2

Keys	Display	Comments
MODE FIX 3	0.000	Show 3 places of decimals
(3 ÷ 8)	0.375	Result of 3/8
e^x	1.455	Answer

Example 3

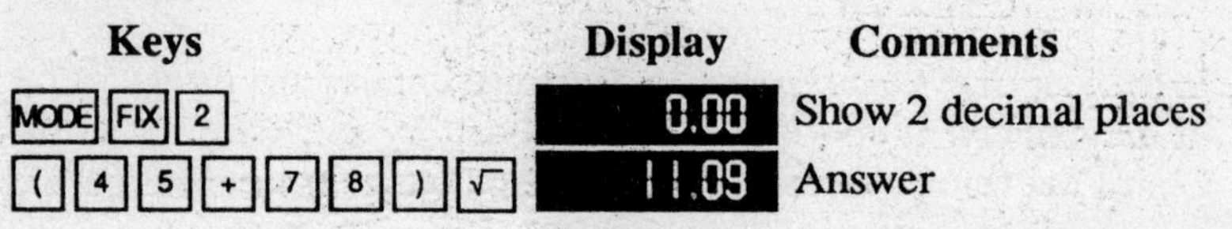

Keys	Display	Comments
MODE FIX 2	0.00	Show 2 decimal places
(4 5 + 7 8) √	11.09	Answer

A more complicated example follows.

The equation for the resonant frequency of a tuned circuit is :

$$f = \frac{\sqrt{\frac{1}{LC} - \frac{R^2}{L^2}}}{2\pi}$$

To calculate this without doing it in two or three separate parts, you must add the brackets which have been left out:

1. Brackets are required around the entire expression under the square root sign.
2. Brackets are essential around each of the two parts of the expression under the square root.

The expression should look like this:

$$f = \frac{\sqrt{\left(\left(\frac{1}{LC}\right) - \left(\frac{R^2}{L^2}\right)\right)}}{2\pi}$$

You can now work out the answer without difficulty.

Example 4

'What is the resonant frequency of a circuit which contains a 27nF capacitor in parallel with a 47mH inductance? The inductance has a resistance of 100 ohms.'

Keys	Display	Comments
MODE SCI 4	0.000	Scientific display
((4 7 EXP +/- 3	47.-03	L
× 2 7 EXP +/- 9)	1.269-09	LC
$^1/_x$	7.880 08	1/LC
- (1 0 0 x^2 ÷	1.000 04	R^2
4 7 EXP +/- 3 x^2)	4.527 06	(R^2/L^2)
) √ ÷ 2 ÷ π =	4.455 03	frequency = 4.455kHz

Chapter 2

Display Control

This chapter shows you how to control the way your calculator displays the results of any calculation. The display can be tailored to your particular requirements, making the machine easier to use.

Reasons for using Display Control

Most Casio owners use their calculators to perform calculations with ordinary decimal numbers. If no attempt is made to control its MODE of operation it will be in MODE NORM. In this mode it will show the maximum number of decimal places and answers will often fill the entire display.

The result is usually that the answers shown have far more places of decimals than required, making the display more confusing than it need be. As a result, you may read the answer incorrectly and you have the job of extracting and rounding the most significant part of the answer before writing it down. By controlling the display mode it can be tailored to your requirements, reducing the chances of error and displaying the rounded number automatically.

Your calculator will still operate to its full accuracy however, since only the display is affected.

FIX – Controlling the Number of Decimal Places Shown

This mode determines the maximum number of figures shown after the decimal point. By pressing [MODE] [FIX] followed by any number from 0 to 9 the number of decimal places displayed is set to this last number. For example, [MODE] [FIX] [4] sets the display to show 4 figures after the decimal point.

Example 1

'A student gets 43 marks out of 65 for a piece of work. What % mark is this, correct to 1 place of decimals?'

Keys	Display	Comments
[MODE] [FIX] [1]	0.0	Show 1 decimal place
[4] [3] [÷] [6] [5] [×] [1] [0] [0] [=]	66.2	Answer

Switching the machine back to normal mode [MODE] [NORM] will display the answer to the limits of the calculator display, i. e. 66.15384615.

Example 2

'Find the Sine of 67 degrees correct to four places of decimals.'

Keys	Display	Comments
[MODE] [FIX] [4]	0.0000	Show 4 places of decimals
[MODE] [DEG]	0.0000	Operate in Degree mode
[6] [7] [SIN]	0.9205	Answer

To display whole numbers only, select 0 places of decimals:

Example 3

'What is the radius, to the nearest metre, of a circle whose area is 5000 square metres?'

Keys	Display	Comments
[MODE] [FIX] [0]	0.	Whole numbers only
[(] [5] [0] [0] [0] [÷] [π] [)] [√]	40.	Answer

SCI – Controlling the Number of Significant Figures Displayed

It is not always practicable to display a fixed number of decimal places. For instance:

'Divide 12 by 2378'

The answer to this is 0.005046257, and entering [MODE] [FIX] [9] will cause it to be displayed. If the calculator is operated to display only two places of decimals, i.e. [MODE] [FIX] [2], it will show an answer of 0.01! This is clearly unsatisfactory as, if this answer is taken at face value, it is about double the correct one!

A better method is to set the number of significant figures using Scientific Mode. Answers are then always displayed in Standard Form with the required number of significant figures. Entering

[MODE][SCI] followed by any number from 0 to 9 will set the number of significant digits shown.

For example, [MODE][SCI][3] will select Scientific mode and show 3 significant digits.

[MODE][SCI][0] will show 10 significant figures.

Example 4

'Divide 12 by 2378, giving the answer to four significant figures'

Keys	Display	Comments
[MODE][SCI][4]	0.000 00	Show 4 sig. figures
[1][2][÷][2][3][7][8][=]	5.046 -03	Answer

If you are using your calculator for scientific or engineering calculations the Scientific mode is usually the best.

ENG – Engineering Mode

Engineering mode is the same as Scientific mode except that the powers of 10 are always in multiples of 3. For example, the speed of light in Scientific mode (MODE SCI 4) is shown as:

2.998 08 metres per second

In Engineering mode, the same number is shown as:

299.8 06 metres per second

Engineering units of measurement are arranged in a hierarchy of 1000s (e.g. millimetres, metres, kilometres etc.), so Engineering mode automatically selects the correct units.

Engineering mode is normally selected by first obtaining the answer required and then pressing the [ENG] key.

The display will show the number of digits already selected by [MODE][FIX] or [MODE][SCI].

Example 5

'What value of capacitor has an impedance of 500 ohms at a frequency of 1kHz?'

The calculation to be performed is:

$$\frac{1}{2 \times \pi \times 1000 \times 500}$$

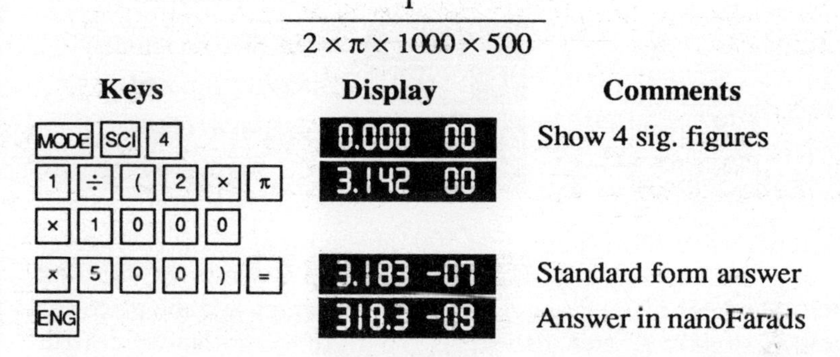

Some models of Casio scientific calculators have a special ENG mode which is selected by MODE ENG. The display will then contain the word ENG.

The number of digits shown is pre-selected by MODE FIX or MODE SCI as explained above. Using MODE ENG with these machines shows the power of 10 as a letter. E.g. 1000s are shown as k, 1,000,000s as M etc.

For example, 23 million is normally displayed as:

23. 06 after using the ENG key

and 23. M after using the MODE ENG keys.

To exit from ENG mode press MODE ENG keys again.

RND – Rounding Off Answers

FIX and SCI mode control how answers are displayed, but do not affect the accuracy with which the calculator works internally. The RND key used in conjunction with FIX and SCI rounds off answers not only in the display but also inside the calculator so that further calculations use the rounded number shown on the screen. A typical application is in the following calculation of costs:

Example 6

'The cost per square centimetre of sheet metal stampings is 3p. Find the cost of a circular stamping whose diameter is 15 cm, and calculate the cost of 500 of these.'

Keys	Display	Comments
MODE FIX 2	0.00	Show pounds and pence
π × 1 5 x^2	225.00	Diameter squared
÷ 4 × . 0 3 =	5.30	Cost/unit to nearest p
× 5 0 0 =	2650.72	Cost of 500 ????

A rapid calculation shows that 500 units costing £5.30 each gives a total of £2650 exactly. The extra 72 pence is due to the fact that the calculator is operating to maximum accuracy and the price per unit is slightly greater than £5.30. To make the arithmetic correct from the accountant's point of view the price per unit must be rounded internally to match what you see on the screen before multiplying to find the cost of 500. Here is the calculation again:

Keys	Display	Comments
MODE FIX 2	0.00	Show pounds and pence
π × 1 5 x^2	225.00	Diameter squared
÷ 4 × . 0 3 =	5.30	Cost/unit to nearest p
RND	5.30	Round internal answer
× 5 0 0 =	2650.00	Cost of 500.

Chapter 3

Polar to Rectangular P→R & Rectangular to Polar R→P

The P→R and R→P functions are used to solve some types of right-angled triangle problems. They are particularly useful when vectors need to be split into two components at right angles, and when dealing with the reverse situation – putting a vector back together again. They also have many other uses in mechanical and electrical engineering.

What are Polar and Rectangular Coordinates?

These two immensely useful functions complement each other and are therefore dealt with together. All current Casio scientific calculators include these functions. On the keyboard they are marked P→R and R→P respectively. They are intended to solve right-angled triangle problems. Before considering how to use them you must first understand the meaning of the terms *Polar* and *Rectangular* coordinates.

There are two ways of giving the position of one point on a map in relation to another:

First, you can say that it is X km East and Y km North of the reference point. This is the system used when quoting a National Grid reference for a place on the map. In the case of the National Grid, the reference point is somewhere off the Isles of Scilly. This system is called *Rectangular Notation.*

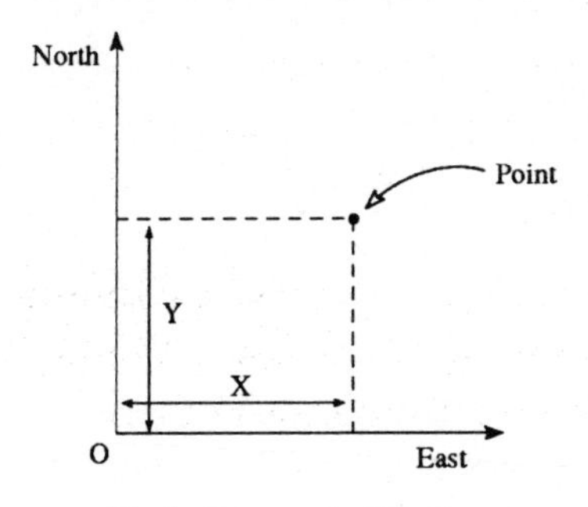

Fig. 1 – Rectangular Notation

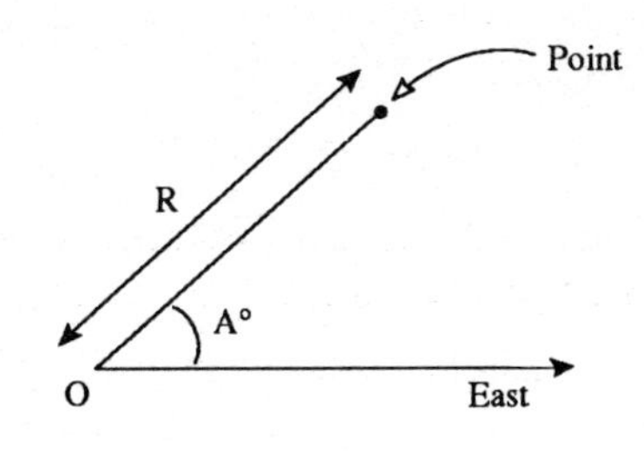

Fig. 2 – Polar Notation

Second, you can say that the same place is R km in a straight line from the reference point at an angle of A degrees anticlockwise from the line pointing East from the reference. This is *Polar Notation.*

Since both methods can refer to the same point it is possible to convert from one system to the other. To anyone familiar with trigonometry it will be obvious that when converting Polar to Rectangular Coordinates:

$$X = R \times Cos\ A$$

and $$Y = R \times Sin\ A$$

And when converting from Polar to Rectangular form that:

$$R^2 = X^2 + Y^2$$

and $$A = arctan\ (X/Y)$$

You are simply solving a right-angled triangle problem whenever you convert from Polar to Rectangular form or vice versa.

[R→P] – Rectangular to Polar Conversion

The [R→P] key provides a swift way of converting from Rectangular coordinates to Polar coordinates without using the equations given above.

The procedure is:

1. Enter the X (horizontal) distance
2. Press the [R→P] key
3. Enter the Y (vertical) distance
4. Press [=]
5. Read the distance R from the display
6. Press the [X↔Y] key
7. Read the angle A from the display

Further use of [X↔Y] will alternately display the distance and the angle (R and A).

Any problem where the two sides of a right-angled triangle are known, and the length of the hypotenuse and one of the angles need to be discovered, can be solved easily using this function.

In all the following examples you can assume that the calculator is set to operate in Degrees and is in *FIX* 2 mode:

Example 1

'Find the length of a ladder, and its angle to the horizontal if, when propped against a wall, its base is 5.18 feet away from the wall and it reaches 19.32 feet up the wall.'

Obviously, in this case, $X = 5.18$ and $Y = 19.32$. The length of the ladder will be R and the angle A.

Keys	Display	Comments
5 . 1 8	5.18	X – horizontal distance
R→P	5.18	
1 9 . 3 2	19.32	Y – vertical distancc
=	20.00	R – ladder length
X↔Y	74.99	A – angle in degrees

There is an enormous range of applications for this function, most of which are fairly obvious. There are, however, a number of other applications (in electrical theory for example) which are not immediately obvious.

Example 2

'Find the impedance and phase angle of a relay coil whose resistance is 4.36 ohms, and whose inductive reactance is 49.81 ohms.'

In this case the resistance can be regarded as X and the inductive reactance as Y. The sequence is:

Keys	Display	Comments
4 . 3 6	4.36	X – resistance
R→P	4.36	
4 9 . 8 1	49.81	Y – reactance
=	50.00	R – impedance in ohms
X↔Y	85.00	A – degrees phase angle

Example 3

'An aircraft flies due North at an airspeed of 600 km/hr for two hours and during this time it drifts 100km to the East due to a crosswind. How far is it from its departure point when it lands?'

In this case the 100km travelled East is X and the distance travelled North (2 × 600km) is Y. The final distance from home is R. The angle is not required.

Keys	Display	Comments
1 0 0	100.	X – drift Eastwards
R→P	100.00	
1 2 0 0	1200.	Y – distance Northwards
=	1204.16	R – distance from start

P→R – Polar to Rectangular Conversion

This key is the inverse of the R→P key and converts polar to rectangular coordinates.

The procedure is:

1. Enter the radial distance (R)
2. Press the P→R key
3. Enter the angle (A)
4. Press =
5. Read the distance X from the display
6. Press the X↔Y key
7. Read the distance Y from the display

Further use of X↔Y will alternately display the distance and angle.

Any right-angled triangle problem where the length of the hypotenuse and one other angle are known, is easily solved by the P→R function.

Example 4

'A ladder 15 feet long is propped up against a wall so that the angle between the ladder and the ground is 70 degrees. Calculate the distance between the foot of the ladder and the wall, and the height of the top of the ladder from the ground.'

This is a straightforward Polar to Rectangular conversion. The length of the ladder is *R* and its angle is *A*. The distance from the wall to the base of the ladder is *X*, the height that it reaches up the wall is *Y*. Here is the sequence to solve the problem:

Keys	Display	Comments
1 5	15.	*R* – ladder length
P→R	15.00	
7 0	70.	*A* – angle
=	5.13	*X* – ladder to wall
X↔Y	14.10	*Y* – distance up wall

Further operation of X↔Y will alternately display the wall to ladder distance and the height reached up the wall.

Again, many applications for this function are obvious, but there are many others, particularly in the electrical field.

Example 5

'An inductive coil has an impedance of 500 ohms with a phase angle of 30 degrees. Calculate the resistance and inductive reactance of the coil.'

As electrical students will know, this is another problem concerning a right-angled triangle. The impedance is *R*, the phase angle *A*, the resistance will be *X* and the inductive reactance *Y*.

Keys	Display	Comments
5 0 0	500.	*R* – impedance
P→R	500.00	
3 0	30.	*A* – phase angle
=	433.01	*X* – ohms resistance
X↔Y	250.00	*Y* – ohms reactance

Further operation of X↔Y alternately shows resistance and reactance.

Chapter 4

Angle Functions

There are several units in which angles are measured. Degrees are the most common, but Radians are often used in science and engineering and it is important to know when radians must be used.

Degrees, Radians and Gradients

Most angles are measured in Degrees, and the calculator will operate in this mode unless it is deliberately set to Radians or Gradients. If the calculator is not in Degree mode, [MODE] [DEG] will set it up. The display will then contain the word *DEG* or *D*.

Some machines now retain their *MODE* when switched off. In that case it is important that you check and, if necessary, change the *MODE* of operation before using the calculator.

The scientific unit of angle is the Radian. If a circle is drawn and a distance equal to its radius is measured around the circumference, the two ends of this arc make an angle of 1 radian at the centre. The total circumference of a circle is its radius multiplied by 2π, so it follows that 2π Radians = 360 Degrees. 1 Radian = 57 Degrees approximately. [MODE] [RAD] selects Radians, when the word *RAD* or *R* will appear in the display. Once selected, angles are assumed to be in Radians, which is important when using the three Trigonometric functions [SIN], [COS] or [TAN], their inverses or other functions which make use of these. Examples of such functions are [P→R] and [R→P].

Few people have ever heard of Gradients – apart from their appearance on scientific calculators. There are 100 Gradients in a right angle and therefore 400 Gradients in a circle. The calculator can be switched to operate in Gradients with [MODE] [GRA]. The display will then contain the word *GRA* or *G*. I know of no common applications for Gradients.

Don't ever leave your calculator in Gradient mode. If you do and you use it for calculations involving angles, thinking that it's in degrees mode, all your answers will be wrong. The trouble is that they probably won't be 'out' enough for you to realise what has happened.

Example 1

'Find the Tangent of 23 degrees correct to 4 decimal places'

Keys	Display	Comments
MODE FIX 4	0.0000	4 decimal places
MODE DEG	0.0000	Degrees mode
2 3 TAN	0.4245	Answer

Example 2

'What angle, in degrees to the nearest degree, has a Cosine of 0.9?'

Keys	Display	Comments
MODE FIX 0	0.	Integer only
MODE DEG	0.	Degrees mode
. 9 COS^{-1}	26.	Answer

Example 3

'What is the Sine of 1.5 Radians, correct to 3 decimal places?'

Keys	Display	Comments
MODE FIX 3	0.000	3 decimal places
MODE RAD	0.000	Radians mode
1 . 5 SIN	0.997	Answer

Example 4

'Find the angle in Radians whose Cosine is 0.6, correct to 4 places of decimals.'

Keys	Display	Comments
MODE FIX 4	0.0000	4 decimal places
MODE RAD	0.0000	Radians mode
. 6 COS^{-1}	0.9273	Answer

There are some situations where Radians *must* be used because the formula assumes their use.The commonest example is when using equations for simple harmonic motion – there are many of these in science and engineering.

Example 5

'The voltage output v of an ac generator has the equation:

$$v = 50 \sin(314t)$$

Calculate the voltage when t = 0.0015 to 2 decimal places'

Keys	Display	Comments
[MODE] [FIX] [2]	0.00	2 decimal places
[MODE] [RAD]	0.00	Radians mode
[(] [3] [1] [4] [×]		
[.] [0] [0] [1] [5] [)]	0.47	Angle in radians
[SIN] [×] [5] [0] [=]	22.69	Answer in volts

One model of scientific calculator has three special *SHIFT* modes which are labelled [o], [r] and [g] (i.e. degrees, radians and gradients). The only difference between these modes and the ones described above is that any number in the display is automatically converted to its equivalent value in the new unit when changing units.

For example, suppose [MODE] [DEG] has been keyed in to the calculator with the number 90 displayed. In order to enter radian mode, [MODE] [RAD] can be keyed in but it will leave the 90 unchanged.

However, if [MODE] [SHIFT] [r] is keyed instead, radians mode will be entered and the 90 will change to 1.571, which is the equivalent of 90 degrees in radians.

Chapter 5

Binary, Octal and Hexadecimal

Anyone studying computing needs to understand and use number bases other than the usual base of 10. Finding the value of a number in another base is a frequent problem. Dealing with negative numbers is particularly awkward. This chapter shows you how to convert negative and positive numbers from one base to another.

Number Bases

All digital computers operate in Binary – the number system containing only the digits 0 and 1 (i.e. base 2). Because Binary numbers are very confusing to the human eye, computer numbers are often displayed in Octal (base 8) or Hexadecimal (base 16). The digits in Octal run from 0 to 7, in Hexadecimal they run from 0 to 9 and then A to F, where A = 10, B = 11, C = 12, D = 13, E = 14 and F = 15. In general, memory and register contents in mainframe computers are shown in Octal, whilst microcomputers show register contents in Hexadecimal.

Operations with these three number bases using Casio scientific calculators are subject to certain restrictions which are:

a All results are whole numbers.

b Numbers or parts of numbers less than 1 are lost.

c The maximum size of numbers is restricted.

d Negative numbers are not preceded by a – sign, but are shown in 2's complement form.

Subject to the above restrictions, numbers in Binary, Octal or Hexadecimal can be added, subtracted, multiplied and divided, using brackets if necessary to change the order in which calculations are performed.

Conversion between Decimal, Binary, Octal and Hexadecimal

The calculator should be switched to the appropriate mode of operation by [MODE] [BASE-N].

Direct conversions between the different systems are then obtained by pressing [DEC], [BIN], [HEX] or [OCT] with or without the [SHIFT] key as appropriate. Alternatively your calculator may have separate Modes for Binary, Octal, Decimal and Hexadecimal.

The following examples assume that [MODE] [BASE-N] is used:

Example 1

'What are the Binary, Octal, and Hexadecimal equivalents of the Decimal number 150?'

Keys	Display	Comments
[MODE] [BASE-N] [DEC]	0.	Base-n mode, decimal
[1] [5] [0]	150.	Decimal number
[BIN]	10010110.	Binary equivalent
[OCT]	226.	Octal equivalent
[HEX]	96.	Hexadecimal equivalent

Negative numbers in Binary, Octal and Hexadecimal are displayed in 2s complement form. This method uses the leftmost binary digit as an indication of sign, leaving the rest of the number to give the value. The reason for doing this is that computers use binary (i.e. the symbols 0 and 1 and nothing else.) The only way to indicate the sign is therefore to use one of the digits for this purpose. A *1* indicates a negative number, A *0* a positive number.

You are probably used to dealing with 8 and 16 – bit binary numbers or their hexadecimal equivalents. Scientific calculators usually work in 10 or 12 binary digits, the same number of Octal digits and 8 Hexadecimal digits – depending on the width of the display.

For this reason negative Binary, Octal and Hexadecimal numbers will usually be shown with more digits than expected. The next example shows how to deal with this on the assumption that your calculator shows a maximum of 10 Binary and Octal digits, and 8 Hexadecimal digits.

Example 2

'What are the Binary, Octal and Hexadecimal equivalents of –150?'

Keys	Display	Comments
MODE BASE-N DEC	0.	Base-n, Dec
1 5 0 +/–	-150.	Decimal number
BIN	1101101010.	Binary equivalent
OCT	7777777552.	Octal equivalent
HEX	FFFFFF6A.	Hexadecimal equivalent

In the case of the Octal and Hexadecimal numbers shown above you will need to trim off the leading digits to suit your particular purpose. For example a 16-bit Hexadecimal number contains 4 digits, so –150 = FF6A in Hex. A 15-bit Octal number contains 5 digits and –150 = 77552 in Octal.

If you want to convert a 2's complement negative number in Binary, Octal or Hex back to decimal you will need to pad out the left hand side of the number with 1s, 7s or Fs respectively to use all the available digit positions.

The next example shows this.

Example 3

*'Find the decimal equivalent of the Hexadecimal number FA. Assume that it is **not** in 2's complement form':*

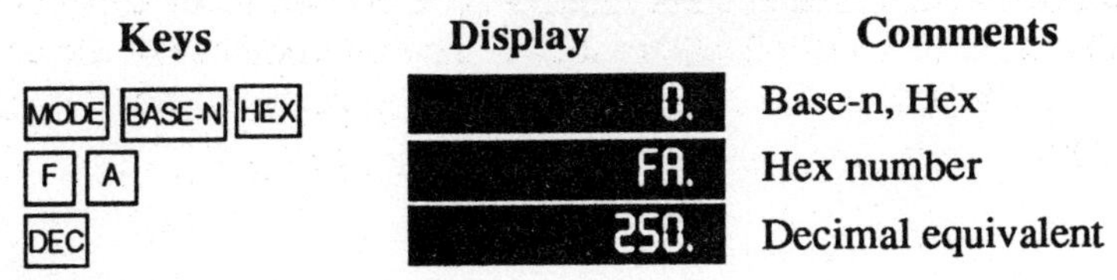

Keys	Display	Comments
MODE BASE-N HEX	0.	Base-n, Hex
F A	FA.	Hex number
DEC	250.	Decimal equivalent

Example 4

'Find the decimal equivalent of the 8-bit hexadecimal number FA. It is in 2's complement form'.

The most significant digit is F. Hexadecimal numbers in 2's complement form whose most significant digit is in the range 8 to F are negative, therefore extra leading Fs must be added to fill the display. The largest number of Hexadecimal digits that can be displayed on, for example, the Casio model fx–570c is eight, the number must therefore be entered as FFFFFFFA.

Keys	Display	Comments
[MODE] [BASE-N] [HEX]	0.	Base-n, Hex
[F] [F] [F] [F] [F] [F] [F] [A]	FFFFFFFA.	Hex number
[DEC]	-6.	Decimal equivalent

NEG – Negating a number

The [NEG] key changes the sign of any displayed number.

If the calculator is in Decimal mode then pressing the [NEG] key simply toggles the minus sign on/off. In *HEX*, *OCT* or *BIN* mode, the number is changed from positive to negative or negative to positive using the 2's complement system.

Example 5

'Change the sign of the Hexadecimal number 3D2, and find the decimal value of the result.'

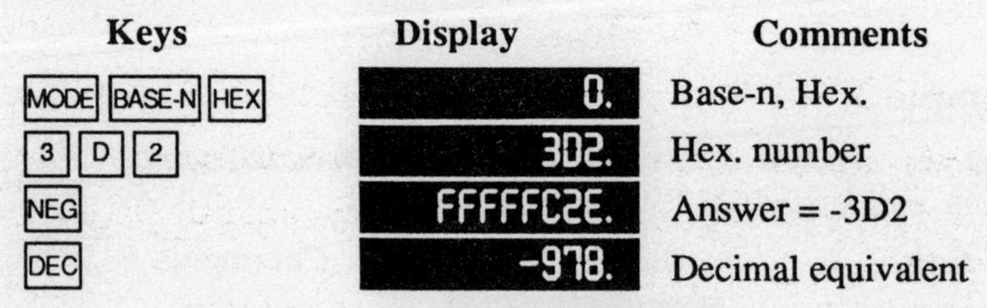

Keys	Display	Comments
[MODE] [BASE-N] [HEX]	0.	Base-n, Hex.
[3] [D] [2]	3D2.	Hex. number
[NEG]	FFFFFC2E.	Answer = -3D2
[DEC]	-978.	Decimal equivalent

Chapter 6

Logic Functions

This chapter explains the operation of the logic functions and how to understand the answers your calculator gives.

NOT, AND, OR, XOR and XNOR

These functions can only be used when the calculator is in *BASE–N* mode, and they should not be confused with other functions. In particular, [NOT] and [NEG] perform different functions; [AND] has no connection with ordinary arithmetic addition.

All of these functions except [NOT] produce a result by means of a logical operation between two numbers. In every case the answer can only be understood by considering both numbers in their Binary form.

Logical operations are carried out on a column-by-column basis. There is no carry from one column to another, unlike ordinary arithmetic operations. Calculations can be chained together like ordinary mathematical operations – brackets being used, where necessary, to change priorities.

All of the examples shown assume that your calculator can display a maximum of 10 Binary,10 Octal and 8 Hexadecimal digits.

NOT

The *NOT* operator exchanges 0s for 1s and 1s for 0s in any binary number: i.e. *NOT* 1 = 0 and *NOT* 0 = 1. It follows from this that:

	Binary
[NOT]	0100010101
[=]	1011101010

– assuming that your calculator has a 10 digit binary display.

This is also true if these two numbers are in Decimal, Octal or Hexadecimal:

	Decimal		Octal		Hexadecimal
[NOT]	277	[NOT]	425	[NOT]	115
[=]	–278	[=]	7777777352	[=]	FFFFFEEA

In the case of the Decimal answer, the Binary result is assumed to be in 2's complement form, in this case giving a negative answer.

In the Octal and Hexadecimal cases you need to remember that both of the original numbers were preceded by a string of 0s which were suppressed by the calculator display.These are treated as Binary zeroes and inverted to 1s when the conversion is made. These 1s are displayed as 7s in Octal and Fs in Hexadecimal.

AND

The [AND] operator *AND*s the binary digits in a pair of Binary numbers column by column to produce the answer. The rules are:

0 *AND* 0 = 0
0 *AND* 1 = 0
1 *AND* 0 = 0
1 *AND* 1 = 1

Hence:

	Binary
	10110100
[AND]	101000
[=]	100000

The same operation gives the equivalent result in Decimal, Octal and Hexadecimal:

	Decimal		**Octal**		**Hexadecimal**
	180		264		B4
[AND]	40	[AND]	50	[AND]	28
[=]	32	[=]	40	[=]	20

OR

The [OR] operator *OR*s the binary digits in a pair of Binary numbers column by column to produce the answer. The rules are:

0 *OR* 0 = 0
0 *OR* 1 = 1
1 *OR* 0 = 1
1 *OR* 1 = 1

Hence:

	Binary
	10110100
OR	101000
=	10111100

The same operation gives the equivalent result in Decimal, Octal and Hexadecimal:

	Decimal		Octal		Hexadecimal
	180		264		B4
OR	40	OR	50	OR	28
=	188	=	274	=	BC

XOR

This operator performs an *Exclusive–OR* between two Binary digits in the same column. The rules are:

0 *XOR* 0 = 0
0 *XOR* 1 = 1
1 *XOR* 0 = 1
1 *XOR* 1 = 0

Hence:

	Binary
	10110100
XOR	101000
=	10011100

The Decimal, Octal and Hexadecimal equivalents produce the same result:

	Decimal		Octal		Hexadecimal
	180		264		B4
XOR	40	XOR	50	XOR	28
=	156	=	234	=	9C

XNOR

This operator performs an *Exclusive–NOR* between pairs of Binary digits, column by column. The result is the inverse of the *XOR* answer in every case. The rules are:

0 *XNOR* 0 = 1
0 *XNOR* 1 = 0
1 *XNOR* 0 = 0
1 *XNOR* 1 = 1

Hence:

	Binary
	10110100
XNOR	101000
=	1101100011

The Decimal, Octal and Hexadecimal equivalents produce the same result:

	Decimal		**Octal**		**Hexadecimal**
	180		264		B4
XNOR	40	XNOR	50	XNOR	28
=	–157	=	7777777543	=	FFFFFF63

As you will see from the table above, if these operations are attempted on the calculator they produce the rather unexpected results shown. The reason is that the *XNOR* is simply an *XOR* and a *NOT*. Leading 0s in the answer have been inverted to 1s, giving a negative result and loading the left-hand end of the Octal and Hexadecimal answers with 7s and Fs respectively.

Applications of Logic Functions

Most Scientific Computer Languages and all Assembly Languages include logic functions in their instruction set. All these functions operate on a bit-by-bit basis as explained above. The main uses for these functions are in manipulating and/or reading bits within numbers. There are no direct applications for these functions on a non-programmable calculator.

Chapter 7
Standard Deviation

This chapter shows you how easy it is to calculate the average of a set of numbers and both forms of standard deviation. Examples are given together with a guide to error correction.

What is Standard Deviation?

When you calculate the average (mean) of a set of figures this only gives half the story. Another important factor is the spread or scatter of the individual figures around the mean. Standard deviation is a measure of scatter.

For instance, manufacturers of powerful modern drugs are concerned not only with the average weight of each tablet but also with the scatter of the individual values of weight, because of the obvious danger of an underdose or overdose. There are many other examples where the scatter of a set of values is as important as their average.

There are two different kinds of Standard Deviation. One is called Population Standard Deviation for which the symbol is σ_n. This is the number calculated when all the values of interest are known, e.g. the income of every single person in a town.

The other kind of Standard Deviation is the Sample Standard Deviation. This is the measure used when the values from which it is calculated are a sample from a population. For instance, if every tenth citizen of a town is asked about his income then the figures obtained are used to calculate the Sample Standard Deviation. Its symbol is σ_{n-1}.

If you enter a set of numbers into the calculator you will find that the Sample Standard Deviation is always slightly higher than the Population Standard Deviation.

Calculating the Standard Deviation and Mean

The calculator must first be put in the *SD* (Standard Deviation) MODE.

Most current Casio calculators have a special [MODE] key which is used to select *SD*.

Sometimes there is no [MODE] key, but there is a key marked [SD].

It can usually be selected by first pressing [INV] or [SHIFT]. The display will then show *SD*. The numbers are then typed in one at a time, followed by [DATA] in each case.

One current model uses a key marked [XD] instead of [DATA].

The examples below assume that your machine has a [DATA] key.

When the calculator is in *SD* mode it automatically calculates both types of standard deviation together with the mean of all the numbers entered ($\bar{x}$), their sum (Σx), the sum of their squares (Σx^2) and the number of entries (n).

These six items are usually shifted functions of other keys. If this is so they will be identified in the same colour as the [SHIFT] key next to the six appropriate keys.

On calculators equipped with constant memories (i.e. with [Kin] and [Kout] keys) they share only three keys i.e. each key gives two of these functions as well as its normal one.

The items shown in *SHIFT* colour can be accessed by pressing [SHIFT] first.

The other three by pressing [Kout] first. In other words, [Kout] behaves like an additional shift key in these circumstances.

Example 1

'Ten electric lamps were taken as samples from a production line and life tested. The lives of the lamps in hours were:

1100, 1045, 1067, 997, 1000, 1150, 1120, 1105, 1098, 1070.

Find the mean life of the lamps, the sample standard deviation of the life and the sum of the lives of all the lamps.'

Keys	Display	Comments
MODE FIX 0	0.	Display whole numbers
MODE SD	0.	Set to SD mode
KAC	0.	Clear all memories
1 1 0 0	1100.	1st number
DATA	1100.	Store number
1 0 4 5	1045.	2nd number
DATA	1045.	Store number

and continue through list until...

Keys	Display	Comments
1 0 7 0	1070.	10th number
DATA	1070.	Store it
n	10.	Check 10 figures entered
$\bar{x}$	1075.	Mean life
$x\sigma_{n-1}$	50.	Sample standard deviation
Σx	10752.	Sum of all the lives.

Example 2

'An exam paper is marked out of 20. The results gained by 15 candidates were:

4 got 14 marks, 3 got 15 marks, 1 got 20 marks, 3 got 13 marks, 1 got 10 marks and 3 got 16 marks.

What was the average (mean) mark and the population standard deviation of the marks?'

Keys	Display	Comments
KAC	0.	Clear All Memories
MODE SD	0.	Standard Deviation Mode
MODE FIX 2	0.00	2 places of decimals
1 4	14.	
DATA DATA DATA DATA	14.00	4 lots of 14 marks
1 5	15.	

Keys	Display	Comments
DATA DATA DATA	15.00	3 lots of 15 marks
2 0	20.	
DATA	20.00	20 marks once
1 3	13.	
DATA DATA DATA	13.00	3 lots of 13 marks
1 0	10.	
DATA	10.00	10 marks once
1 6	16.	
DATA DATA DATA	16.00	3 lots of 16 marks
n	15.00	Check no. of entries
$\bar{x}$	14.53	Average mark
$x\sigma_n$	2.09	Population std. deviation

Correcting Mistakes

When entering a large set of numbers there is always the chance of making a mistake. It is not usually necessary to start the entire calculation again if you enter a wrong number. How you get round the problem depends on the kind of mistake you have made. There are several possibilities; you may have typed a wrong number without pressing DATA, you may have pressed DATA as well as getting the number wrong, and you may have realised that you entered a wrong number mixed in with a set of right ones several numbers back. In every case there is a way to correct the mistake without starting again *providing you know what the mistake is.*

Error	How to deal with it
Typed in wrong number without using *DATA*.	Press *C*. Then re-enter correct value. Press *DATA*.
Typed in wrong number followed by *DATA*, but number still on display.	Press *DEL*. Then re-enter correct number followed by *DATA*.
Wrong number entered several items back.	Re-enter *wrong* number, then use *DEL*. This will remove number. Then enter correct number followed by *DATA*.

Chapter 8

Linear Regression LR

Linear Regression is particularly useful if you need to draw the best straight line through a set of points on a graph, need to make estimates based on a set of figures which form an approximately straight line, or to find out how closely two sets of figures relate to each other. This Chapter explains Linear Regression and gives a number of examples to show how versatile this function is. There is also a short section on correcting mistakes.

What is Linear Regression?

The easiest way to understand Linear Regression is to consider the problem of drawing the best possible straight line through a set of points plotted on a sheet of graph paper. Many sets of experimental results form what is supposed to be a straight line when plotted as a graph.

The standard equation for a straight line graph is of the form:

$$y = A + Bx$$

where A is the value of y when x = 0 (intercept on y axis) and B is the 'slope' of the graph.

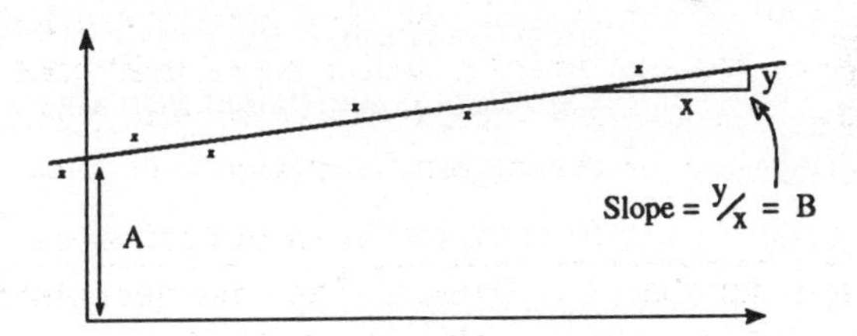

With your calculator in *LR* mode and the data entered, it will give you the slope (B) and intercept (A) of the best straight line through the points. In addition, as a bonus, much other useful information is available – see the table below.

Almost all these extra items are self-explanatory apart from the correlation coefficient and estimator functions. These need more explanation.

The coefficient of correlation (*r*) is a figure which expresses how accurately the points fit the straight line. If $r = 1$, all the points fall precisely on a straight line which slopes upwards from left to right (a positive slope). If $r = -1$ the same is true, but the line slopes the other way (a negative slope). If $r = 0$ there is no connection whatever between the two sets of data – they will simply appear as a random set of points if any attempt is made to plot them. In general, if the absolute value of *r* is 0.8 or greater it suggests a reasonably tight fit of the points around the line.

The estimator functions $\hat{x}$ and $\hat{y}$ enable you to estimate the value of *x* for a given value of *y*, or the value of *y* for a given value of *x* based on the best straight line through the points. For example, to find what *y* value corresponds to a known value of *x*, you key in the value of *x* and press [$\hat{y}$]. This is an immensely useful way of estimating the position of any point on the line. See the examples later in this chapter for typical applications.

Special Key Functions

Apart from the data entry keys, there are a number of keys with special functions when in LR mode. They are:

Keys	Function
[A]	Intercept of graph on y axis
[B]	Slope of graph
[r]	Correlation coefficient (see above)
[$\bar{x}$] [$\bar{y}$]	Averages of x and y
[Σx] [Σy]	Sum of x and y values
[Σxy]	Sum of x.y terms
[n]	Number of data pairs entered
[Σx^2] [Σy^2]	Sum of x squared and y squared
[$x\sigma_n$] [$y\sigma_n$]	Population standard deviations
[$x\sigma_{n-1}$] [$y\sigma_{n-1}$]	Sample standard deviations
[$\hat{x}$] [$\hat{y}$]	Estimated values of x and y

Where two of these functions are written next to another key, the function *not in SHIFT colour* is obtained by pressing [Kout] first. [Kout] acts as a *SHIFT* key in this case. Some of the latest models make this clear by adding this information to the Mode list printed underneath the calculator display window.

Don't confuse the Linear Regression *A* and *B* keys with the two other keys also marked *A* and *B* and which form part of the set *A* to *F* used for hexadecimal number entry.

Applications of Linear Regression

This function has many useful applications. Any set of experimental data which is expected to approximate to a straight line graph can make use of linear regression. It can:

1. Help to put the best possible line through the points.
2. Estimate where points will fall outside the graph (extrapolation).
3. Estimate where points will fall within the graph (interpolation).

Data Entry

Data entry for linear regression is more complex than the other functions. Points you should note are:

1. Data is entered in pairs, i.e. the two numbers relating to the same point on the graph are entered separated by a key marked [X_DY_D], followed by [DATA]. For example, if x = 3 and y = 4:

   [3] [X_DY_D] [4] [DATA]

2. The two numbers of each pair must always be entered in the same order, first x then y.

Example 1

'The length of a coil spring was measured when a series of different weights were suspended from it. The results obtained were:

Length in cm	*15.0*	*17.1*	*19.0*	*20.9*	*22.8*	*24.9*	*27.0*	*29.1*
Weight in gm	*50*	*100*	*150*	*200*	*250*	*300*	*350*	*400*

Estimate the length of the spring when it is unloaded, carries a load of 500 gm, or a load of 220 gm, and the weight required to make the spring 22 cm long. What is the spring rate (the change in length per gm)?'

You need to decide whether the lengths or weights are *x*. The others will then be *y*. If the weights are *x*, the weight must be entered first every time you key in a pair of figures. To estimate the length of the spring for a given weight you need to use [ŷ] after typing in the weight.

Keys	Display	Comments
[MODE] [FIX] [2]	0.00	2 decimal places
[MODE] [LR]	0.00	Linear regression mode
[KAC]	0.00	Clear all memories
[5] [0] [X_DY_D]	50.00	First weight
[1] [5] [DATA]	15.00	First length
[1] [0] [0] [X_DY_D]	100.00	Second weight
[1] [7] [.] [1] [DATA]	17.10	Second length
[1] [5] [0] [X_DY_D]	150.00	Third weight
[1] [9] [DATA]	19.00	Third length

....continue entering data until...

Keys	Display	Comments
[4] [0] [0] [X_DY_D]	400.00	Final weight
[2] [9] [.] [1] [DATA]	29.10	Final length
[n]	8.00	Check no. of pairs
[A]	12.99	Unloaded length
[5] [0] [0] [ŷ]	32.96	Length with 500 gm load
[2] [2] [0] [ŷ]	21.78	Length with 220 gm load
[2] [2] [x̂]	225.63	Weight required for 22 cm
[B]	0.04	Spring rate (cm per gm)

Example 2

'The sales of dongles have risen steadily year by year. Results for the past six years were:

Sales in thousands	*23.6*	*25.4*	*27.4*	*29.5*	*31.7*	*33.6*
Year	*1983*	*1984*	*1985*	*1986*	*1987*	*1988*

Estimate the sales for 1989 and 1992. In what year should sales reach 40 thousand? What are the average and total sales over the period covered in the table?'

If the year is the first number entered in every data pair it is x.

Keys	Display	Comments
MODE LR	0.	Linear regression mode
MODE FIX 1	0.0	1 place of decimals
KAC	0.0	Clear all memories
1 9 8 3 X_DY_D	1983.0	First year
2 3 . 6 DATA	23.6	First sales figure
1 9 8 4 X_DY_D	1984.0	Second year
2 5 . 4 DATA	25.4	Second sales figure

...continue entering pairs of figures until...

Keys	Display	Comments
1 9 8 8 X_DY_D	1988.0	Final year
3 3 . 6 DATA	33.6	Final sales figure
n	6.0	Check number of pairs
1 9 8 9 $\hat{y}$	35.6	Projected 1989 sales
1 9 9 2 $\hat{y}$	41.7	Projected 1991 sales
4 0 $\hat{x}$	1991.2	40,000 sales year
$\bar{y}$	28.5	Average yearly sales
Σy	171.2	Total sales 1983 - 88

Example 3

'A class of students takes two written tests in the same subject. The marks (out of 20) were:

Student No.	*1*	*2*	*3*	*4*	*5*	*6*	*7*	*8*	*9*	*10*
Test 1 mark	*15*	*10*	*8*	*19*	*12*	*9*	*8*	*17*	*15*	*12*
Test 2 mark	*17*	*12*	*9*	*20*	*13*	*9*	*11*	*18*	*18*	*Abs*

Student 10 missed the second test. Estimate the mark this student would have gained in this test. find the average mark gained in each test and the coefficient of correlation between the two sets of test results.'

We will take the Test 2 marks as *x*.

Keys	Display	Comments
[MODE] [LR]	0.	Linear Regression mode
[KAC]	0.0	Clear all memories
[MODE] [FIX] [0]	0.0	Display integers only
[1] [7] [X_DY_D] [1] [5] [DATA]	15.	First pair of results

...continue entering results until...

Keys	Display	Comments
[1] [8] [X_DY_D] [1] [5] [DATA]	15.	Final pair of results
[n]	9.	Check no, of data pairs
[1] [2] [$\hat{x}$]	14.	Estimated mark
[$\bar{x}$]	14.	Average mark in test 2
[$\bar{y}$]	13.	Average mark in test 1
[MODE] [FIX] [2]	12.56	Display 2 decimal places
[r]	0.97	Correlation is good.

Correcting mistakes

When entering a large set of numbers there is always the chance of making a mistake. It is not usually necessary to start the entire calculation again if you enter a wrong number or pair of numbers. How you get round the problem depends on the kind of mistake you have made. There are several possibilities; you may have typed a wrong number or pair of numbers without pressing DATA, you may have pressed DATA as well as getting the numbers wrong, and you may have realised that you entered a wrong pair of numbers mixed in with a set of right ones several entries back. In every case there is a way to correct the mistake without starting again *providing you know what the mistake is.*

Error	How to deal with it
Typed in wrong number without using X_DY_D	Press *C*. then re-enter correct value. Continue entering pair.
Typed in wrong number followed by X_DY_D	Type in correct number, then X_DY_D, then second number of pair, then *DATA*
Second number of pair entered wrongly without pressing *DATA*.	Press *C*, then enter correct number, then press *DATA*.
Wrong pair of numbers entered, including *DATA*. Last number still visible on display.	Press *DEL*, then enter correct pair.
Wrong number pair entered several items back.	Re-enter wrong pair, separated by X_DY_D, followed by *DEL*. This will remove wrong pair. Then enter correct pair of numbers, separated by X_DY_D, followed by *DATA*.

Chapter 9

Factorials, Permutations and Combinations

These three functions are very useful in situations where things have to be arranged in a number of different ways, teams picked from an available set of players, and other problems of a similar nature. This chapter explains what these functions do and how to use them.

Factorials [x!]

The factorial of a number (which must be a positive integer) is that number multiplied by every other integer smaller than itself down to a value of 1.

The factorial of a number is usually indicated by an exclamation sign after the number. So factorial 5 is shown as 5! and:

$$5! = 5 \times 4 \times 3 \times 2 \times 1 = 120$$

This is, for example, the number of ways that 5 books can be arranged on a shelf.

Example 1

'In how many ways can the 11 members of a football team be allocated different positions?'

Keys	Display	Comments
[MODE] [FIX] [0]	0.	Integers only
[1] [1] [x!]	39916800.	Number of ways

Permutations [nPr]

The number of permutations of 2 things out of 10 is the number of possible arrangements of 2 items chosen from 10. For instance the number of ways that 2 books can be chosen from 10 and arranged on a shelf. The answer treats the two ways that each pair of books can be arranged as separate permutations.

The number of permutations of r things chosen from n is usually written nPr.

Example 2

'A shelf holds 8 books. How many possible arrangements of books to fill the shelf are possible if there are a total number of 10 books to choose from?'

Keys	Display	Comments
MODE FIX 0	0.	Integers only
1 0 nPr 8	1814400.	Answer

Combinations nCr

The number of combinations of 2 things taken from 10 is the number of ways that 2 items could be chosen from 10. The answer is different to nPr because the ways in which those two things can be arranged after being chosen is ignored. The number of combinations of r things chosen from n is written nCr. Since there are only two ways of arranging two things, the answer to 10P2 is half that of 10C2.

Example 3

'Out of a pool of 13 players, how many different teams of 11 can be chosen?'

Keys	Display	Comments
MODE FIX 0	0.	Integers only
1 3 nCr 1 1	78.	Answer

Chapter 10

Random Numbers

This chapter shows you how to use the random number function to simulate a throw of dice and to choose samples for test or inspection purposes.

Why Random Numbers?

Random numbers have two main uses:

1. In sampling problems
2. In games, to simulate a roll of dice for example

The second application is usually confined to games played against a computer or programmable calculator.

Sampling is however a very common activity in industry, particularly when random inspection of goods is carried out.

The Random Function

Using the [RAN#] key gives a pseudo-random result in the range 0.000 to 0.999. For sampling purposes you normally need a whole number answer. For example, to simulate the throw of a single die, answers 1, 2, 3, 4, 5 and 6 are the only valid possibilities.

To do this:

a The calculator must be set to *MODE FIX 0*

b The result of using *RAN#* must be mutiplied by the *range* of the expected final answer + 1. In this case (6 – 1) + 1 gives a multiplier of 6.

c Finally 0.5 must be added to allow for the fact that the lowest possible answer is 1, not 0.

Everywhere else in this book you should get identical answers to the examples. If you try the examples below it is most unlikely that this will happen because you will be getting *random* answers.

Example 1

'Produce a random pair of numbers to simulate the throw of a pair of ordinary dice.'

Keys	Display	Comments
MODE FIX 0	0.	Integers only!
RAN#	0.621	
× 6 + . 5 =	4.	1st die
RAN#	0.804	
× 6 + . 5 =	5.	2nd die

Example 2

'Out of every 200 artillary shells, 5 are test fired. Assuming the shells in the batch are numbered 1 to 200, make a random choice of the five to be tested.'

In this case the range of possible answers + 1 = 200. The lowest possible answer is 1, so you must add 0.5 to get the final result. Obviously the same number could occur twice and while this doesn't matter when throwing dice, you can't test the same shell more than once! If this happens you simply keep trying until you get an unused number.

Keys	Display	Comments
MODE FIX 0	0.	Integers only!
RAN#	0.021	Random fraction
× 2 0 0 + . 5 =	5.	1st choice
RAN#	0.881	Random fraction
× 2 0 0 + . 5 =	177.	2nd choice
RAN#	0.876	
× 2 0 0 + . 5 =	176.	3rd choice
RAN#	0.415	
× 2 0 0 + . 5 =	84.	4th choice
RAN#	0.114	
× 2 0 0 + . 5 =	23.	Final choice

Chapter 11

Hyperbolic Functions

Unless you are doing quite advanced work in Maths, Science, or Engineering it is unlikely that you will need to use hyperbolic functions. This chapter explains how to use your calculator to obtain these functions and gives an example of their use.

What are Hyperbolic Functions?

Hyperbolic Sines, Cosines and Tangents have the abbreviations Sinh, Cosh and Tanh in most Mathematics text books. They have properties in common with ordinary Sines, Cosines and Tangents – hence the names.

They can be evaluated by any suitably equipped Casio calculator by pressing the HYP key before either *SIN*, *COS* or *TAN*. For their definitions and properties refer to any advanced Mathematics text.

These functions have a large number of applications in advanced science and engineering – for example, the expression y = cosh (x) is the equation for the curve of a uniform and frictionless chain suspended between two points.

Example

'A wire is suspended between two points which are at the same height above the ground. The points are 100m apart horizontally and the height y of any point of the wire is given by the equation:

$$y = 200\cosh(.005x)$$

where x is the horizontal distance of any point from the lowest part of the hanging wire. The distance along the wire from its centre to a point x metres away horizontally is given by 's' where:

$$s = 200\sinh(.005x)$$

Find the minimum distance between the wire and the ground, the sag of the wire at its centre and the length of the wire'.

Keys	Display	Comments
[MODE] [FIX] [2]	0.00	2 decimal places
[0]	0.	x = 0 at centre
[HYP] [COS]	1.00	Cosh of 0
[×] [2] [0] [0] [=]	200.00	Minimum height from ground
[5] [0]	50.	Centre to end dist.
[×] [.] [0] [0] [5] [=]	0.25	(.005x) at supports
[HYP] [COS] [×] [2] [0] [0] [=]	206.28	Height at supports
[-] [2] [0] [0] [=]	6.28	Sag at centre
[5] [0] [×] [.] [0] [0] [5] [=]	0.25	(.005x) at supports
[HYP] [SIN] [×] [2] [0] [0] [=]	50.25	Centre to supports
[×] [2] [=]	101.04	Total length of wire

Chapter 12
Log Functions

Before electronic calculators became available, precision calculations were carried out by looking up logarithms (logs) in mathematical tables. However, logs do have other uses and appear in some Engineering and Scientific formulae. Logs are therefore still required in some circumstances. This chapter gives examples of their use.

Common and Natural Logs

Ordinary or common logs use 10 as a base. The inverse of a log is its antilog where:

$$\text{Antilog}_{10}(x) = 10^x$$

Naperian or Natural logs use '*e*' (2.71828...) as a base. The inverse of a Natural log is its Antilog where:

$$\text{Antilog}_e(x) = e^x$$

Common logs are obtained by using the [LOG] key, natural logs by using the [LN] key.

Common antilogs use the [10^x] key and natural antilogs the [e^x] key.

Example 1

'An electronic amplifier gives an output of 15 watts when the input is 0.007 watts. Calculate its gain in decibels (dB).'

This is an example of the use of common logs in a formula. The formula is:

$$\text{gain in dB} = 10 \log_{10}\left(\frac{\text{Power out}}{\text{Power in}}\right)$$

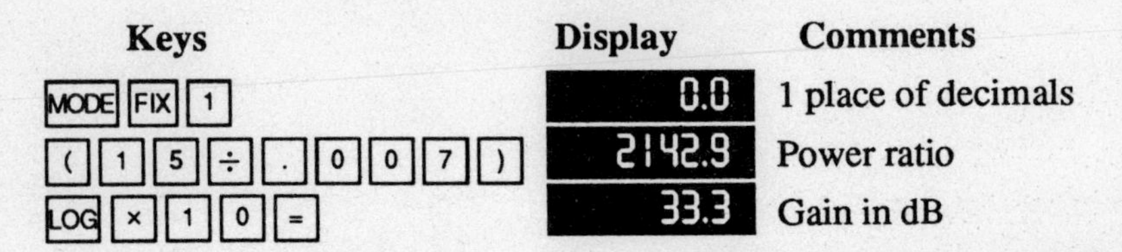

Keys	Display	Comments
MODE FIX 1	0.0	1 place of decimals
(1 5 ÷ . 0 0 7)	2142.9	Power ratio
LOG × 1 0 =	33.3	Gain in dB

Example 2

'A furnace has a maximum temperature rise of 400 degrees Celsius and a time constant of 30 minutes. How long will it take for the temperature to rise by 350 degrees to the nearest minute?'

This uses natural logs, the formula for the time *t* is:

$$t = T\log_e\left(\frac{M}{M-c}\right)$$

where *T* is the time constant, *M* the maximum temperature rise, and c the desired temperature rise.

Keys	Display	Comments
MODE FIX 0	0.	Display integers
4 0 0 ÷ (4 0 0 −	400.	
3 5 0) =	8.	*M/(M-c)*
LN × 3 0 =	62.	Time to nearest min.

Chapter 13

Using Memory

This chapter shows you how to make use of the various memory functions on your calculator. Sensible use of your calculator's memory will save you time and trouble as well as improving the accuracy of calculations.

Calculator Memories

All Casio scientific calculators have at least one memory (*M*) which can be used in a variety of ways. Memory *M* can be used not only to store numbers, but to keep a running total by making use of the *M+* and *M–* facilities. If any number other than zero is stored in *M* the display shows the letter '*M*'.

Some models have a set of other memories (called 'constant' memories) in addition to *M*. If your calculator has these, you will find keys marked [Kin] and [Kout] on your machine. The Constant Memories can only be used for number storage, but because there are six of them they are still very useful. No indication is given if the constant memories are in use, but *X*<->*K* can be used to check the latter. See below.

Memory '*M*' Keys

The various memory key functions are listed below:

Function	Explanation
[Min]	Writes displayed number to memory *M*. Destroys any previous memory contents.
[MR]	Copies memory *M* contents to display.
[M+]	Adds display number to memory contents.
[M–]	Subtracts display number from memory contents.
[M↔X]	Exchanges memory and display contents. Can be used to verify memory contents without destroying display contents.

'Constant' Memory Keys

[Kout] (1 to 6)	Copies constant memory contents to display. This key must be used with number keys 1 to 6 to access the correct constant memory. E.g. [Kout] [3].
[Kin] (1 to 6)	Writes displayed number to constant memory 1 to 6, e.g. [Kin] [4]. Destroys previous contents of memory addressed.
[X↔K] (1 to 6)	Exchanges display and contents of addressed constant memory. E.g. [X↔K] [2]. This is a way of retrieving constant memory contents without losing the display contents. One use is to inspect the contents of the various constant memories during a calculation.
[KAC]	Clear all constant memories to 0.

Using Memory 'M'

This is a very useful feature provided on all modern Casio scientific calculators and most older models as well. The latest machines retain the contents of *M* regardless of whether the power is on or not. This is particularly useful where a calculation is interrupted and the machine switches itself off.

Example 1

'You have been given £500 to spend. Use the calculator to store the amount of money remaining as it is spent.'

This is an example where the ability of memory *M* to retain its contents and carry out arithmetic is very useful.

Keys	Display	Comments
[5] [0] [0] [Min]	500.	Store £500 in *M*
[4] [×] [2] [3] [.] [6] [5] [=]	94.6	Cost of four shirts
[M–]	94.6	Subtract from total
[1] [3] [9] [.] [9] [9]	139.99	Cost of a suit
[M–]	139.99	Subtract from total

(Calculator switches itself off in the interval)

Keys	Display	Comments
ON	0.	Switch on machine
MR	265.41	Check remaining money
4 5	45.	Cost of personal stereo
M-	45.	Subtract from total
3 × 4 . 2 0 =	12.6	3 packs of stereo tapes
M-	12.6	Subtract from total
MR	207.81	Total remaining.

Using the Constant Memory

These are completely independent of memory *M*, described above, but memory arithmetic of the kind shown in the previous example is not possible. Machines equipped with constant memories often have a register exchange feature, making it possible to verify constant memory contents without affecting the progress of a calculation.

Constant memory is particularly useful where you have a whole series of calculations of the same type to perform.

Example 2

'A triangle has two sides whose lengths are 12cm and 15cm respectively. Find the length of the third side if the angle between the other two sides is 30, 45, 60, 75, 90, 105 or 120 degrees.'

If the length of the third side is a, the lengths of the other two sides are b and c, and the included angle is A:

$$a^2 = b^2 + c^2 - 2bcCos(A)$$

You need to calculate a as the angle A varies. b and c are fixed. The obvious way to speed up calculations is to store:

$b^2 + c^2$ in K1

and $2bc$ in K2

Keys	Display	Comments
[1] [2] [x^2] [+] [1] [5] [x^2] [=]	369.	$b^2 + c^2$
[Kin] [1]	369.	Store in K1
[2] [×] [1] [2] [×] [1] [5] [=]	360.	*2bc*
[Kin] [2]	360.	Store in K2
[MODE] [FIX] [2]	360.00	Show two decimal places
[Kout] [1] [−] [Kout] [2] [×]	360.00	Recover both stores
[3] [0] [COS] [=] [√]	7.57	Length at angle 30 deg.
[Kout] [1] [−] [Kout] [2] [×]	360.00	Recover both stores
[4] [5] [COS] [=] [√]	10.70	Length at angle 45 deg.

continuing through the angles until...

Keys	Display	Comments
[Kout] [1] [−] [Kout] [2] [×]	360.00	Recover both stores
[1] [2] [0] [COS] [=] [√]	23.43	Length at angle 120 deg.

Chapter 14

Fractions

Due to metrication, fractions are not used now as much as they used to be. They still have their uses however, and a calculator that can work in fractions is very useful. This chapter shows you how to use the fraction handling ability of Casio scientific calculators.

Entering Fractions

To enter a fraction into the calculator you need to use the [ab/c] key to separate the different parts of the fraction from each other. Normally a special symbol like a reversed L in small capitals appears in the display to separate the parts of the fraction. One current model separates the parts with a '/'. Thus, twelve and a half is entered:

Keys	Display	Comments
[1] [2] [ab/c] [1] [ab/c] [2]	12⅃1⅃2.	Twelve and a half

To display the equivalent vulgar fraction (i.e. with no whole number at the front), use the shifted function [d/c]:

Key	Display	Comment
[d/c]	25⅃2.	25/2

Repeated use of the [d/c] key toggles the fraction between its proper and vulgar form.

Once the fraction has been completely entered, using the [ab/c] key again toggles the number between its decimal equivalent and fractional value.

Key	Display	Comments
[ab/c]	12.5	Decimal equivalent

Fractional numbers can be added, subtracted, multiplied and divided using brackets where required. The answer will be given as a fraction of the lowest possible denominator. The memory keys can be used in the normal way and recalling memory will return a fractional number. The ordinary function keys (*SIN* for example) can be used with fractional numbers if required.

Example 1

'A bolt measures three and five-sixteenths of an inch long. How long is a bolt half this length?'

Keys	Display	Comments
3 ab/c 5 ab/c 1 6	3 ⌟5 ⌟16.	Length of bolt
÷ 2 =	1 ⌟21 ⌟32.	$1^{21}/_{32}$

Example 2

'Subtract three-sevenths from three-quarters and multiply the result by five. Give the answer as a vulgar fraction and as a decimal number to three places of decimals.'

Brackets must be used to ensure that the subtraction is carried out before the multiplication.

Keys	Display	Comments
MODE FIX 3	0.000	Show 3 decimal places
(3 ab/c 4 −	3 ⌟4	3/4
3 ab/c 7)	9 ⌟28	Result of subtraction
× 5 =	1 ⌟17 ⌟28	Proper fraction answer
d/c	45 ⌟28	Vulgar fraction answer
ab/c	1.607	Decimal answer

Chapter 15

Complex Numbers

If you use complex numbers, a calculator that will handle them is a great time-saver. One current Casio scientific calculator will do this – your machine will have a *CMPLX* mode in this case. This chapter shows you how to use your calculator to carry out complex number arithmetic easily.

i and j

Complex numbers consist of two parts, the real part and the complex part. For example:

$$5 + 3i$$

Casio calculators use this notation as it is mathematically correct, although electrical engineers use 'j' instead of 'i'. The majority of common applications for complex numbers are in Electrical network theory. Nearly all the examples given in this chapter therefore relate to problems in electrical networks – but using 'i' instead of 'j'.

Complex Mode

The calculator must first be set in the correct operating mode by pressing the keys [MODE][CMPLX]. The word '*CMPLX*' is then shown at the top of the display.

You leave complex mode by selecting [MODE][COMP], [MODE][BASE-N], etc. Within complex mode the display can be controlled to suit your requirements – see the chapter on Display Control for details. In addition, operation in Degrees, Radians or Gradients is possible – see also the chapter on Angle Functions.

All four ordinary mathematical operations, +, –, × and ÷ can be carried out with ease using brackets around each complex number and other brackets as required. In addition, powers, roots, reciprocals and conjugates of complex numbers can be calculated at the touch of a button. The independent memory *M* retains all its functions, i.e. complex numbers can be stored, summed, recovered and exchanged with the display as required. The modulus (absolute value) and argument (angle) of any complex number can also be obtained.

Since Complex Numbers consist of two parts the display only shows one of these at any one time. You can switch the display between the two parts by using the [Re↔Im] key.The examples below illustrate this.

Entering a Complex Number

Enter the real part of the number first, then a '+' or '–' sign followed by the complex part of the number, then the letter 'i'. To enter the entire number into the calculator you must then either use the [=] key or close the brackets opened when you started. It is best to use brackets around all complex numbers. The display will then show the real part of the complex number. To see the complex part use [Re↔Im]. To see the real part again use [Re↔Im].

e.g. To enter the number 6 + 3i, the key sequence is:

[(] [6] [+] [3] [i] [)]

Doing Complex Arithmetic

Once you know how to enter complex numbers into your calculator you will find that complex arithmetic is very easy.

Example 1

'A current whose value is (6 – 4i) amps flows through an impedance of (4 + 2i) ohms. What is the resulting volt drop?'

You need to multiply the current by the impedance to find the volt drop.

Keys	Display	Comments
[MODE] [CMPLX]	0.	Set to Complex Mode
[(] [6] [–] [4] [i] [)] [×]	6.	Real part of current
[(] [4] [+] [2] [i] [)]	4.	Real part of impedance
[=]	32.	Real part of voltage
[Re↔Im]	-4.i	Complex part of voltage

The answer is therefore 32 – 4i volts.

Example 2

'Find the admittance of an impedance of (6 - 4i) ohms.'

The admittance is the reciprocal of the impedance. This example shows the advantage of display control. If you leave the machine in *MODE NORM* the answer will fill the display. No one needs 10 digits for each part of the answer so let us display only the first 4 decimal places.

Keys	Display	Comments
MODE CMPLX	0.	Set to complex mode
MODE FIX 4	0.0000	Show 4 decimal places
(6 - 4 i)	6.0000	Real part of impedance
1/x	0.1154	Real part of admittance
Re↔Im	0.0769i	Complex part of admittance

The answer is therefore 0.1154 + 0.0769i.

Example 3

'An electrical circuit consists of 2 parallel branches whose impedances are (4 + 3i) and (3 - i) ohms respectively. What is the overall impedance of the circuit in complex form and ordinary form? What is the phase angle of the impedance?'

You will find that the easiest way to do this is to calculate the reciprocal of each impedance and sum them together in memory. You then recall memory and take the reciprocal of this. This is exactly the way you would calculate the value of parallel resistances – except you are working in complex numbers.

Keys	Display	Comments
MODE FIX 4	0.0000	4 dec. places shown
MODE CMPLX	0.0000	Complex mode
(4 + 3 i)	4.0000	Real part of 1st impedance
1/x Min	0.1600	Real part of 1st admittance
(3 - i)	3.0000	Real part of 2nd impedance
1/x M+	0.3000	Real part of 2nd admittance

Both admittances are now summed in M

Key	Display	Description
MR	0.4600	Real part of total admittance
1/x	2.1698	Real part of total impedance
Re↔Im	0.0943i	Complex part of total impedance

Overall impedance in complex form is 2.1698 + 0.0943i ohms

Key	Display	Description
|z|	2.1719	Total impedance in ohms
arg	2.4896	Phase angle in degrees